To imagination, because we'll always need it.

ISBN-13: 978-0615935546
ISBN-10: 0615935540

AF882680

Robot Story

www.lt-automation.com
PH 305 914 5083
FAX 775 637 6825
Miami, FL, USA

Table of Contents

Famous TV Robots

The following Robots are famous because they have appeared regularly in TV series.

We have tried to give you the name of the first one (person or company) who developed the concept or name of the robot. Some characters are based on a book series so, in this case the writer of the books is the creator of the robots.

In some other cases, the robot is assigned to a company or industrial designer, then this last one is considered the robot creator.

The Jetsons

1962

Rosie

ABC, Hanna Barbera

Rosie is a Robot Maid in the Jetsons Family's home. She was initially hired as the Jetson's housekeeper, but very soon she became part of the family.

She is able to prepare delicious pastries and other dishes loved by Mr. George Jetson's boss.

She has no legs. Instead she has a single wheel which allows her to move very fast.

Although Rosie is an older model, the family never replaced her with a new robot.

Astroboy

1963

Astroboy

ABC, Osamu Tezuka Fuji

Astroboy is a robot created by a renowned Scientist named Dr. Tenma, to fill the place of a son who died in an accident.

Astroboy has lots of powers. He uses his powers to protect MetroCity, which is the city where humans and robots live together.

Futurama

1999

Bender

Fox, Matt Groening

Bender is a robot capable of bending metals. He was made in Mexico.

His complete name is Bender Bending Rodriguez.

Bender behaves inappropriately. He likes to tell lies. He also likes to gamble, smoke, and drink. He needs alcohol as a fuel to work.

Invader Zim

2001

Gir

Nickelodeon, Johnen Vasquez

GIR is a robot grabbed out of the trash and assembled with spare parts to help an alien named Zim.

He is not very smart.

He likes to eat junk food, but mainly Tacos.

He wears a green disguise which makes him look like a dog, but he does a lot of things dogs don't do.

My life as a teenage Robot

2003

X-J9

Nickelodeon, Rob Renzetti

The robot X-J9, also known as Jenny, was created by Dr Wakeman to protect the Earth.

She can express emotions such as happiness and sadness.

Jenny has lots of weapons and devices.

In addition to the normal life she wants to live with he classmates, she has to fight several villains.

This page has been intentionally left blank

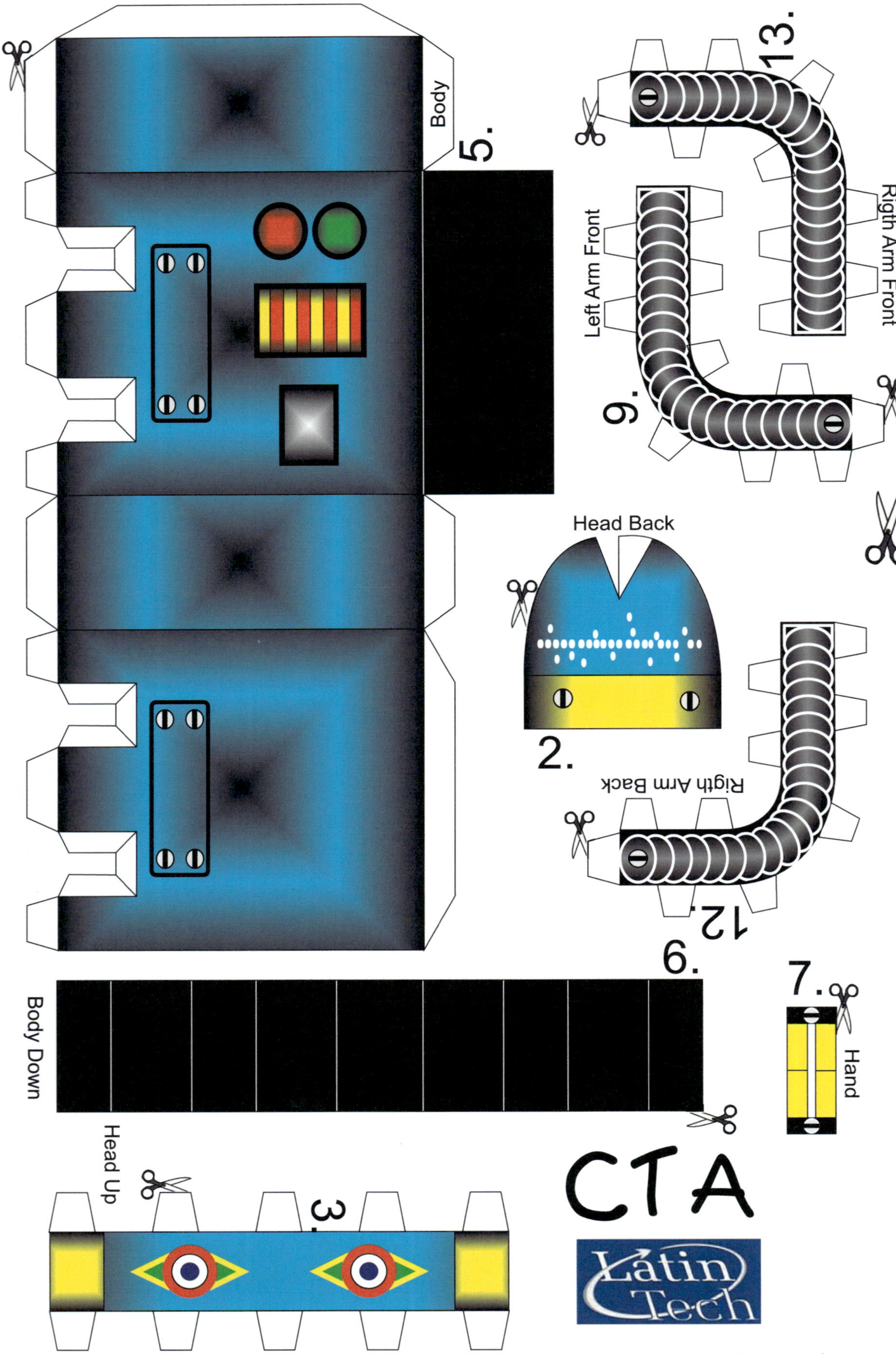
Body
5.
13.
Rigth Arm Front
Left Arm Front
9.
Head Back
2.
Rigth Arm Back
12.
6.
7.
Hand
Body Down
Head Up
3.
CTA
Latin Tech

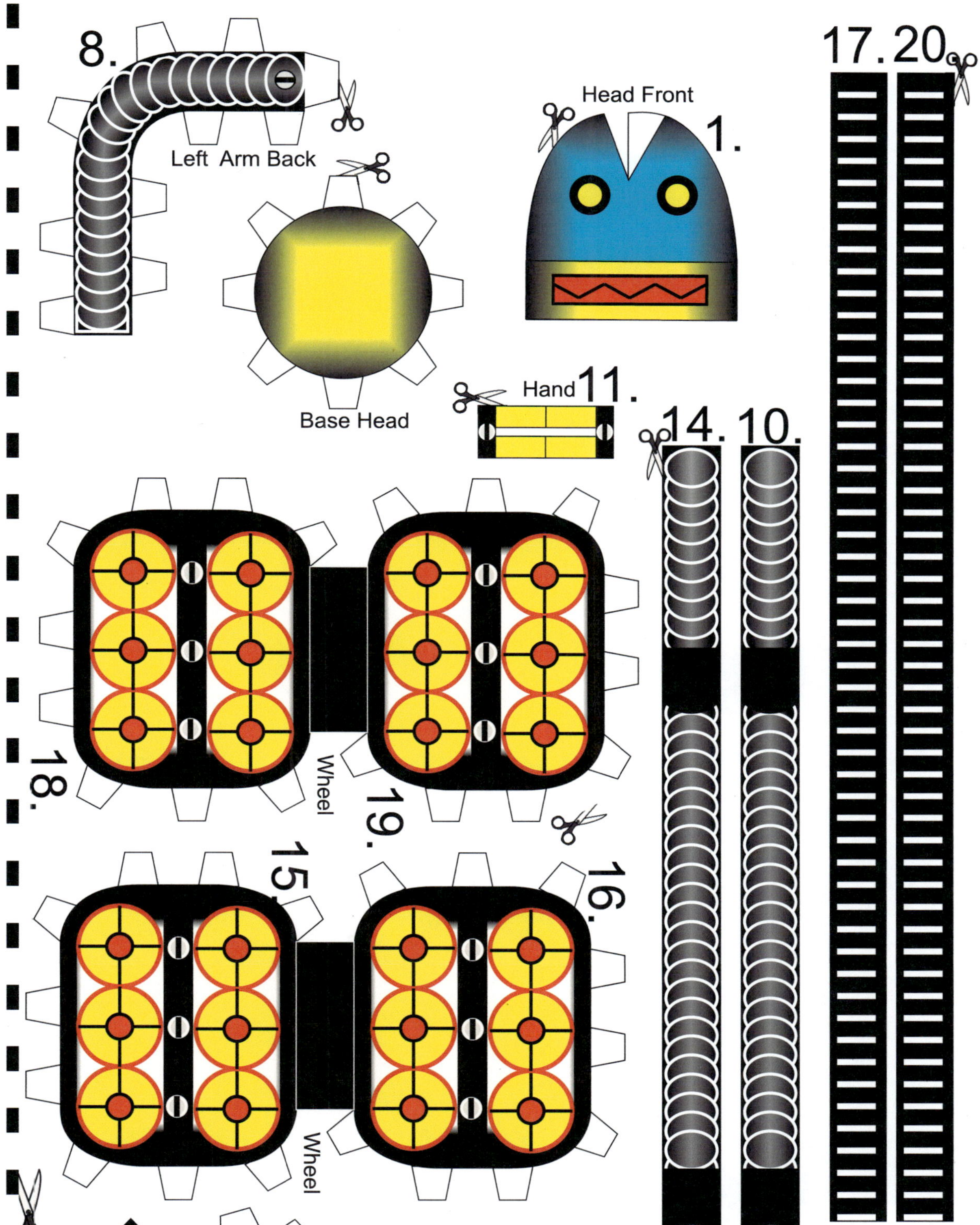

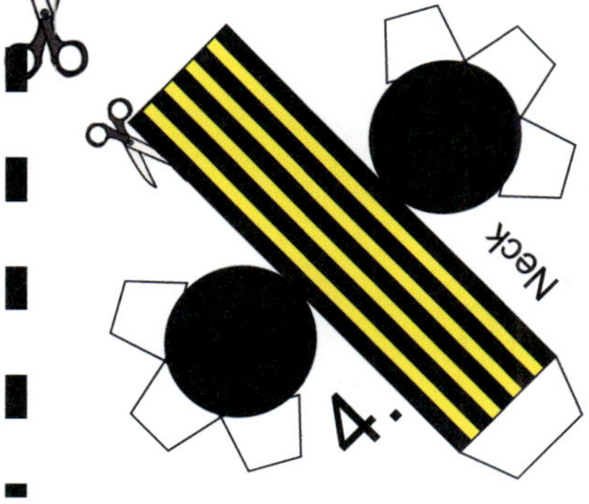

To order more full color papercraft robots or books contact sales@latin-tech.net

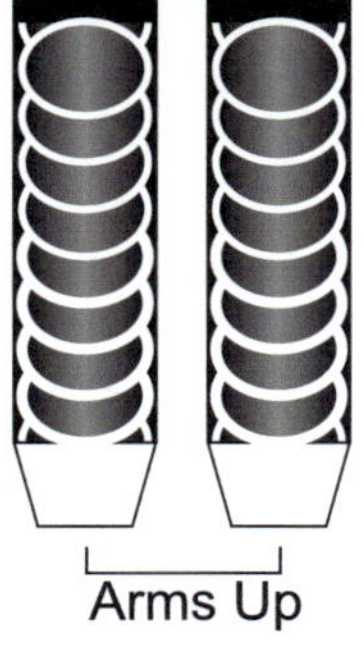

If you require help send an e-mail to sales@latin-tech.net

This page has been intentionally left blank

Jimmy Neutron

2006

Goddard

Nickelodeon

Jhon A. Davis, Keith Alcorn
Steve Oedekerk

This robot dog was created by Jimmy Neutron, a boy genius, who is a very active inventor.

Not all of Jimmy's inventions work.

Goddard is one of Jimmy's successful inventions. He is also Jimmy's best friend.

Goddard has a lot of technical capabilities: he has a screen on his chest and a video projector in his eyes. He can transform himself into a hover bike. He can also explode and reassemble himself.

Thrasher and Blastus Robotomy

2010

Thrasher and Blastus

Cartoon Network

Michael Buckleavy
Joe Deasy

Thrasher and Blastus are two teenage robots attending high school in the planet Insanus.

In this planet, every robot must be a killer and having feelings is not a good thing.

Thrasher is a tall robot who is in love with a very popular high school robot named Maimy.

Blastus is a fat robot who does things without much thinking. He just wants to be popular.

Doctor Who

1963

Dalek

Terry Nation and Raymond Cusick

The Daleks are mutants protected by a mechanical casing made of Dalkanium. They come from the planet Skaro.They have no feelings or remorse when trying to eliminate all life forms and always say "EXTERMINATE" when they appear at any place.

The Daleks have several special weapons to destroy things through a death ray, and a telescopic arm which can be mainly used to read minds, to interface to other technological devices or to measure anyone´s intelligence.

Doctor Who

1966

Cyberman

Kit Pedler and Gerry Davies

The Cybermen are part human and part robots so they are also known as cyborgs. They are from the planet Mondas, which in the past was a sister planet of our beautiful planet earth.

The cybermen were like human beings, but they started a process of replacing parts of their bodies with mechanical systems.

These cyborgs have no emotions since they think it´s a weakness symbol.

They want to take humans to their planet and turn them into Cybermen.

The 2006 redesign of the Cybermen

Lost in the Space

1968

B9

CBS, Irwin Allen

The Model B9 is an environmental robot made to help the Robinson family aboard the space ship Jupiter 2.

The robot is armed and capable of solving problems the family has to face. His best friend is William, the youngest child aboard.

He can move through wheels, talk, laugh and he can also make calculations. He uses his sensors to detect problems and danger.

Logan´s Run

1977

Rem

CBS, William F. Nolan

REM is a 200 years old android robot. He is a very intelligent robot who repairs and maintains other robots.

Most of the humans in Stone City died, but the robot servant still continued to operate.

Some of the robots captured Logan and Jessica. They are two humans who are running away from a place where they have to die when they reach 30 years old.

REM rescued them and together they continued to look for a place named Sanctuary.

Buck Rogers in 25th Century

1979

Twiki

NBC, Glen A. Larson

This small android is an "Ambuquad." He is a special type of robot that works in space mines.

Twiki speaks our language, but makes a especial sound. When he is doing something and talking to other robots. he sounds like: "Biddi-Biddi-Biddi."

Twiki is the Doctor Theopolis' assistant. This robot is a computer assembled on a Disk.

Since Doctor Theopolis can't move, Twiki carries him on his chest using his strong robotic arms.

Small Wonder

1985

Vicky

20th Century Fox, Howard Leeds

The name of this android was created using the initial letters from Voice Input Child Identicant: VICI. Everybody likes to call her Vicki.

She was created by the father of the Lawson's family. He is an engineer who works for the Robotronics company. The family decided to keep her as a secret but they adopted her as a member of the family.

Vicki gets power from an internal atomic battery. She is very strong and runs superfast. She looks like a 10 year old girl, however she is not able to show any emotions.

Stark Trek

1987

DATA

CBS, Gene Roddenberry

Data is an android with very powerful computer capabilities because he has a Positronic brain. He is interested in understanding the human behavior.

He is able to feel and sense. With the help of an integrated circuit, he can develop emotions. He is part of the crew aboard the Enterprise.

Do you remember the following words?

"Space: the final frontier. These are the voyages of the starship Enterprise. Its continuing mission: to explore strange new worlds, to seek out new life and new civilizations, to boldly go where no one has gone before".

Vocabulary

Alien: Someone out of the earth.

Assembled: To fit together the parts or pieces.

Bend: That can be curved.

Boldly: Doing something with no fear.

Casing: An external cover.

Disguise: To change the external appearance.

Environmental: Relating to the environment or the nature around.

Exterminate: To kill or destroy al.l

Happiness: The feeling when you are happy.

Hover Bike: This is a bike that instead of wheels to touch the ground it uses air.

Inappropiately: Not right, Not proper.

Industrial: That is made of work in any factory or industry.

Interface: Interconnection between systems, equipment, concepts, or human beings.

Invention: Something new that nobody has seen or done before.

Inventor: The person who makes new things previously unknown.

Junk food: Food that is not healthy.

Popular: Something or someone that everybody knows, likes or appreciates.

Positronic Brain: The central computer the controls the robot similar to the human´s brain

Reassemble: Once the parts are separated they can be joined again.

Redesign: To make a new version or design.

Remorse: A feeling of being sorry for doing something bad or wrong in the past.

Sadness: The feeling when you are sad.

Screen: A display to show information.

Servant: Someone who is hired to do household or personal works such as cleaning or cooking.

Spare parts: Parts you use to fix equipment that is damaged.

Video projector: A device to project a beam of light, an image or a movie.

Villains: A cruel or malicious person.

Weapons:
Any instrument used to fight or hunt.

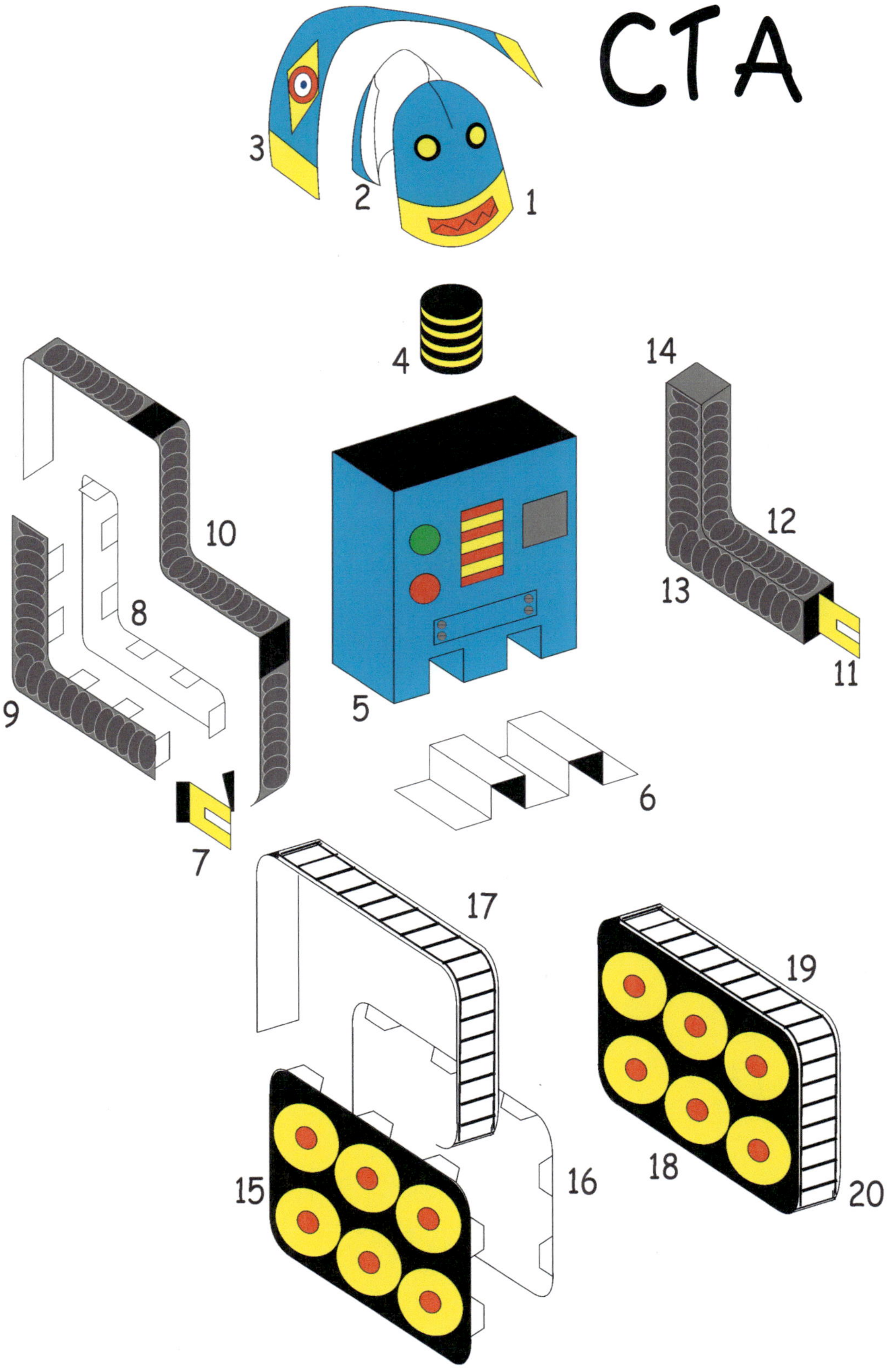

Comming soon:
Online University
www.onlinemediauniversity.com

Made in the USA
Monee, IL
05 November 2026

67377488R00017